GROUNDHOGS

by Martha London

Cody Koala

An Imprint of Pop!
popbooksonline.com

abdobooks.com
Published by Pop!, a division of ABDO, PO Box 398166, Minneapolis, Minnesota 55439. Copyright © 2021 by POP, LLC. International copyrights reserved in all countries. No part of this book may be reproduced in any form without written permission from the publisher. Pop!™ is a trademark and logo of POP, LLC.

Printed in the United States of America, North Mankato, Minnesota

082020
012021

THIS BOOK CONTAINS RECYCLED MATERIALS

Cover Photo: Ron Erwin/All Canada Photos/Alamy
Interior Photos: Ron Erwin/All Canada Photos/Alamy, 1; iStockphoto, 5 (top), 11, 13, 16–17, 19 (bottom right); Shutterstock Images, 5 (bottom right), 5 (bottom left), 8–9; Anne Fournier/Science Source, 6; Leonard Lee Rue III/Science Source, 14; Dominique Braud/Dembinsky Photo Associates/Alamy, 19 (top), 19 (bottom left); Sohns/imageBROKER/Alamy, 21

Editors: Christine Ha and Brienna Rossiter
Series Designer: Sophie Geister-Jones

Library of Congress Control Number: 2019954998
Publisher's Cataloging-in-Publication Data
Names: London, Martha, author.
Title: Groundhogs / by Martha London
Description: Minneapolis, Minnesota : POP!, 2021 | Series: Underground animals | Includes online resources and index.
Identifiers: ISBN 9781532167621 (lib. bdg.) | ISBN 9781532168727 (ebook)
Subjects: LCSH: Groundhog--Juvenile literature. | Woodchuck--Juvenile literature. | Ground squirrels--Juvenile literature. | Rodents--Juvenile literature. | Burrowing animals--Juvenile literature.
Classification: DDC 599.36--dc23

Table of Contents

Chapter 1

Dig, Dig, Dig

Groundhogs are **mammals**. They dig **burrows**. Each burrow has several tunnels and rooms. It also has more than one **entrance**.

Watch a video here!

Groundhogs are found only in North America. They live in parts of Canada and the United States.

Groundhogs dig their burrows in flat land. They often live near trees, streams, and fields.

Groundhogs can dig up to 5 feet (1.5 m) underground.

Chapter 2

Big Teeth

Groundhogs have thick fur. They have short, **bushy** tails. A groundhog can weigh as much as 13 pounds (5.9 kg). Its body can be 20 inches (51 cm) long.

Learn more here!

Groundhogs have large front teeth. Their teeth never stop growing. Groundhogs chew on food, sticks, and roots to wear them down.

Groundhogs eat plants. They eat fruits, vegetables, and nuts.

eye
ear
nose
teeth
claws

Eating and Sleeping

In the summer, groundhogs eat a lot. They build up fat. The fat stores energy. It helps groundhogs get ready for the long winter.

Learn more here!

As the weather gets colder, groundhogs dig more tunnels. They carry grass and leaves into their **burrows**.

Most groundhogs start to **hibernate** in October or November. They sleep in their burrows through the cold winter.

Groundhogs do not eat
during winter. Instead, they
get energy from their fat.

Groundhogs begin to wake
up in the early spring.

When a groundhog hibernates,
its heart slows down. It beats just
five times per minute.

Chapter 4

Small Families

Groundhogs have babies in the spring. A female has up to six babies at one time. The babies are small and **helpless**. Their mother cares for them inside the **burrow**.

Complete an
activity here!

The babies grow and get stronger. They start to leave the burrow. Their mother shows them how to find food.

Groundhogs can live on their own after three months. They dig new burrows. Then they start getting ready for winter.

Making Connections

Text-to-Self

Groundhogs hibernate during winter. What are some activities you usually do during winter?

Text-to-Text

What books have you read about other mammals? What do those animals have in common with groundhogs? How are they different?

Text-to-World

Groundhogs dig burrows. What other types of homes do animals make?

Glossary

burrow – a hole that an animal digs in the ground for shelter.

bushy – thick and fluffy.

entrance – a way to get inside a place.

helpless – not able to take care of oneself.

hibernate – to enter a resting state during winter.

mammal – a type of animal that has hair or fur and feeds milk to its young.

Index

Online Resources

popbooksonline.com

Thanks for reading this Cody Koala book!

Scan this code* and others like it in this book, or visit the website below to make this book pop!

popbooksonline.com/groundhogs

*Scanning QR codes requires a web-enabled smart device with a QR code reader app and a camera.